**Learning Tre
1 2 3**

Adding

By Richard and Nicky Hales

Illustrated by Rebecca Mason

**Geddes+
Grosset**

As you read this book, try to answer the questions, and try to think of some questions of your own.
Write your answers on a piece of paper or in a notebook – not in the book.
There are answers at the end of the book. Try not to look at them before you have had a try. If you find the questions difficult, ask a grown-up or older friend to help you.

First published in hardback 1990
Copyright © Cherrytree Press Ltd 1990

This paperback edition first published 1991 by
Geddes & Grosset Ltd
David Dale House
New Lanark ML11 9DJ

ISBN 1 85534 452 1

Printed and bound in Italy by L.E.G.O. s.p.a., Vicenza

All rights reserved. No part of this publication may be reproduced, stored in a retrieval system, or transmitted, in any form or by any means without the prior permission in writing of the publisher, nor otherwise circulated in any form of binding or cover other than that in which it is published and without a similar condition including this condition being imposed on the subsequent purchaser.

What are these frogs doing?
How many are playing on the climbing frame?
How many are playing on the roundabout?
How many frogs are there altogether?

What are these frogs doing?
How many red frogs are there?
How many yellow frogs?

How many frogs are there altogether?
How many frogs are in tunnels?
How many frogs in a tyre?

How many frogs?
How many frogs are riding bikes?
How many frogs are driving cars?

Are more frogs riding bikes than driving cars?
How many more?
How many frogs are there altogether?

On the lily-pad

Cut out five frogs from cardboard.
Draw two water lilies.

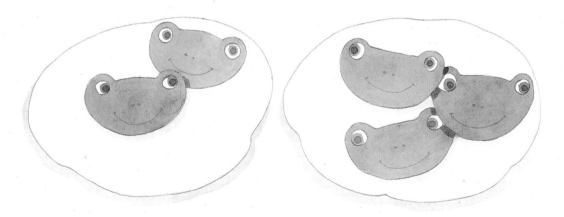

How many ways can you put the frogs on the lilies?

Suppose there were more frogs.
Suppose there were 6 frogs.
How many ways could they go on the lilies?

Dominoes

You can make dominoes like these with pieces of card.

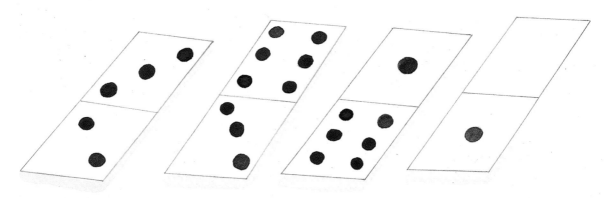

Play dominoes by matching the number of spots. Each player adds a domino to one end of the chain. Keep score by adding the spots in the squares at each end of the chain.

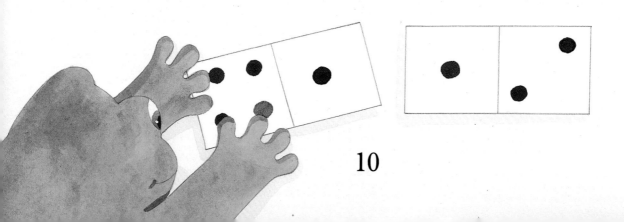

This player has scored 6 (4 spots at one end and 2 at the other).

How many has this player scored?

Set of three numbers

Here is a set of three numbers.

What numbers can you make by adding all the numbers in the set? What numbers can you make by adding some of the numbers in the set? Use each number only once in each sum.

What is the lowest number you can make?

What is the highest?

Can you make all the numbers in between?

Make your own set of three numbers.

Magic squares

Magic squares have numbers that add up the same in all directions.

Can you put these numbers into a square so that each row, column and diagonal adds up to 6?

Make other magic squares and find out more about them.

Use 2,2,2 and 3,3,3 and 4,4,4.

4	2	3
2	3	4
3	4	

Now try 3,3,3 and 4,4,4 and 5,5,5.

Splits

Number 3 can be split up in four different ways.

1 + 1 + 1   1 + 2

2 + 1   3

How many ways can you split up number 4?
Here are two ways.

How many other ways can you find?
How can you split 5 or 6 or 7?

Elevenses

This is a game you can play on your own with a pack of cards.

Take out the picture cards. Shuffle the rest.

Deal out 8 cards face up.
Look for a pair of cards which add up to 11.
When you find a pair, deal a new card on to each.
Carry on until you have used up all the cards.
If you can use them up, it means you have won.

Consecutive numbers

Numbers in order like this are called consecutive numbers.

What happens if you add consecutive numbers?

Try this on a piece of paper:

1 + 2 =
2 + 3 =
3 + 4 =
4 + 5 =
5 + 6 =

Now try this:

1 + 2 + 3 =
2 + 3 + 4 =
3 + 4 + 5 =
4 + 5 + 6 =

Addition squares

Find the missing numbers in these squares.

What questions can you ask?

More Adding

Start an adding notebook. Write the answers to the questions in your notebook.
Do not write in this book.
Someone else might want to use it.

Don't look at the answers until you have tried to work out the answers yourself.

If your answer is not the same as the book's check it. Don't worry if it is still different. Sometimes there is another answer.

When you write down an answer, check that it is right. Turn back to the page and count the things again or add the numbers again.

Make up questions of your own and write the answers.

Count things around you. Count how many chairs at the table at home. How many plates? How many cups? How many saucers? How many cups and saucers altogether?

How many pillows are on each bed? How many windows are there?

Count the number of red cars you see on your way to school. Count the number of trees.

Help with the shopping. Add up the number of things in the shopping basket. What things come in sets? How many bananas are there in a bunch? How can they be split up?

Sort all the items in the shopping basket into sets. You might have a set of tins, a set of packets and a set of things you can't eat. How many are there in each set? Can you sort them in another way?

Work out how much two things in your shopping basket cost.
How much do three things cost?

Help put the shopping away. Do different sets of things go in different cupboards?

1

1 On page 3, how many red frogs are there?
2 How many blue frogs?

3 On page 5, how many frogs are playing on the slide?

4 On pages 6 and 7, how many red frogs are there?
5 How many blue frogs?
6 How many more yellow frogs than green frogs are there?

7 How many red frogs are in the picture below?
8 How many green frogs?
9 How many frogs altogether?

2

10 If there are 3 frogs on one lily-pad and 2 frogs on another lily-pad, how many frogs are there altogether?

11 If there are 4 frogs on one lily-pad and 5 frogs on another, how many frogs are there altogether?

12 What is my score in this game of adding dominoes?

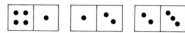

13 What would my friend score if he put down this domino?

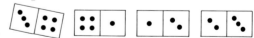

3

14 What answers can you make by adding all or some of these numbers?   1   2   3

One answer is 3 + 2 = 5. What others can you make?

15 How many adding questions can you ask with an answer of 7? Here are some:
1 + 2 + 2 + 2 = 7
5 + 1 + 1 = 7
How many more can you think of?

16 Can you see some ways of making 11 in this game of Elevenses?

6   1   2   8
10   5   3   9

17 What numbers are missing from these addition squares?

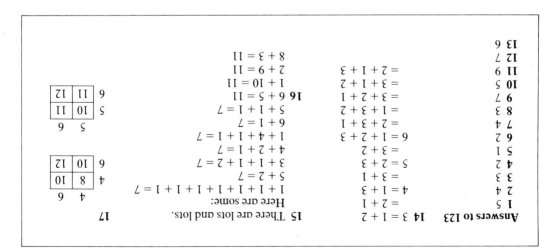

Answers to 123

14 3 = 1 + 2
1 5
2 = 1 + 1
3 = 1 + 3
4 = 1 + 3
   = 2 + 1
5 = 2 + 3
   = 3 + 2
6 = 1 + 2 + 3
   = 1 + 3 + 2
   = 2 + 3 + 1
   = 3 + 1 + 2
   = 3 + 2 + 1

15 There are lots and lots.
Here are some:
1 + 1 + 1 + 1 + 1 + 1 + 1 = 7
5 + 2 = 7
3 + 1 + 1 + 2 = 7
4 + 2 + 1 = 7
1 + 1 + 4 + 1 = 7
6 + 1 = 7
5 + 1 + 1 = 7

16 9 + 5 = 11
1 + 10 = 11
2 + 9 = 11
8 + 3 = 11

17

	4	6
4	8	10
6	10	12

	5	6
5	10	11
6	11	12

23

Index

1 2 3 22, 23
addition squares 19
answers 21, 23, 24
bikes 6, 7
cars 6, 7
check your answers 21
climbing frame 3
consecutive numbers 18

count things 21
dominoes 10, 11
elevenses 17
how many? 21
magic squares 14, 15
make up questions 20, 21
notebook 21

on the lily-pad 8, 9
questions 20, 22, 23
roundabout 3
set of numbers 12, 13
shopping 21
splits 16
what questions? 20

Answers

Page 3
5 frogs on climbing frame
4 frogs on roundabout

Pages 4 & 5
5 yellow frogs
6 red frogs
11 frogs altogether
2 frogs in tunnels
1 frog in a tyre

Pages 6 & 7
5 frogs riding bikes
4 frogs driving cars
1 more riding a bike
9 frogs altogether

Pages 8 & 9
Five frogs   Six frogs
0 and 5   0 and 6
1   4   1   5
2   3   2   4
3   2   3   3
4   1   4   2
5   0   5   1
              6   0

Page 11
⋮⋮ + • = 6

Pages 12 & 13
2 + 3 = 5,
3 + 2 = 5,
2 + 4 = 6,
4 + 2 = 6,
3 + 4 = 7,
4 + 3 = 7,
4 + 3 + 2 = 9,
2 + 3 + 4 = 9,
3 + 4 + 2 = 9,
4 + 2 + 3 = 9,
2 + 4 + 3 = 9,
3 + 2 + 4 = 9

Pages 14 & 15

1	3	2
3	2	1
2	1	3

Each row, column and diagonal add up to 6.

2	4	3
4	3	2
3	2	4

Each row, column and diagonal add up to 9.

3	5	4
5	4	3
4	3	5

Each row, column and diagonal add up to 12.

Page 16
[4]
1 + 1 + 2
1 + 2 + 1
2 + 1 + 1
1 + 3
3 + 1
2 + 2
1 + 1 + 1 + 1
4

[5]
1 + 1 + 1 + 1 + 1
2 + 2 + 1
1 + 2 + 2
2 + 1 + 2
3 + 1 + 1
1 + 3 + 1
1 + 1 + 3
3 + 2
2 + 3
1 + 1 + 1 + 2
1 + 1 + 2 + 1
1 + 2 + 1 + 1
2 + 1 + 1 + 1
1 + 4
4 + 1
5

Page 18
1 + 2 = 3
2 + 3 = 5
3 + 4 = 7
4 + 5 = 9
5 + 6 = 11

In this pattern 2 is added on to each answer.

1 + 2 + 3 = 6
2 + 3 + 4 = 9
3 + 4 + 5 = 12
4 + 5 + 6 = 15

In this pattern 3 is added on each time.

Page 19
In addition squares the diagonal answers are the same.

	3	4
3	6	7
4	7	8

	1	3
1	2	4
3	4	6

	2	4
2	4	6
4	6	8

	5	6
5	10	11
6	11	12

	6	8
6	12	14
8	14	16